S0-AUK-129

# GENETICALLY ENGINEERED FOODS

# OTHER TITLES IN THE SCIENCE ON THE EDGE SERIES INCLUDE:

Cancer Treatments
Forensics
High-Tech Weapons
Nanotechnology
Predicting Natural Disasters
Space Travel

# GENETICALLY ENGINEERED FOODS

WRITTEN BY

## KAREN E. BLEDSOE

**BLACKBIRCH PRESS**

*An imprint of Thomson Gale, a part of The Thomson Corporation*

THOMSON
™
GALE

Detroit • New York • San Francisco • San Diego • New Haven, Conn.
Waterville, Maine • London • Munich

© 2006 Thomson Gale, a part of the Thomson Corporation.

Thomson and Star Logo are trademarks and Gale and Blackbirch Press are registered trademarks used herein under license.

*For more information, contact*
Blackbirch Press
27500 Drake Rd.
Farmington Hills, MI 48331-3535
Or you can visit our Internet site at http://www.gale.com

ALL RIGHTS RESERVED
No part of this work covered by the copyright hereon may be reproduced or used in any form or by any means—graphic, electronic, or mechanical, including photocopying, recording, taping, Web distribution or information storage retrieval systems—without the written permission of the publisher.

Every effort has been made to trace the owners of copyrighted material.

Picture credits: Tek Images/Photo Researchers, Inc., cover; © AJ/IRRI/CORBIS, 29, 37; Alan and Linda Detrick/Photo Researchers, Inc., 16; AP/Wide World Photos, 36, 38; Bloomberg News/Landov, 32; Chris Knapton/Photo Researchers, Inc., 23, 34; © Clouds Hill Imaging Ltd./CORBIS, 19; © CORBIS, 112; Daniel Pepper/Getty Images, 41; Dr. Jeremy Burgess/Photo Researchers, Inc., 26; Dr. Linda Stannard/Photo Researchers, Inc., 30; © Tom Stewart/CORBIS, 40; James King-Holmes/Photo Researchers, Inc., 15; © Jim Richardson/CORBIS, 6; Martyn F. Chillmaid/Photo Researchers, Inc., 25; Mauro Fermariello/Photo Researchers, Inc., 25, 28; Patrick Pleul/DPA/Landov, 43; Photos.com, 11, 13; U.S. Department of Agriculture, 8, 24, 42; © Underwood & Underwood/CORBIS, 18; USDA plant image gallery, 17

## LIBRARY OF CONGRESS CATALOGING-IN-PUBLICATION DATA

Bledsoe, Karen E.
    Genetically engineered foods / by Karen E. Bledsoe.
        p. cm. — (Science on the edge)
    Includes bibliographical references and index.
    ISBN 1-4103-0602-X (hard cover : alk. paper)
    1. Genetically modified foods. 2. Crops—Genetic engineering. 3. Agricultural biotechnology. I. Title. II. Series.

    TP248.65.F66B65 2005
    664—dc22                                                    2005007649

Printed in the United States of America
10 9 8 7 6 5 4 3 2 1

# INTRODUCTION

For thousands of years, humans have been changing the genetics of their food crops and livestock through selective breeding. Farmers learned long ago to save seeds from the best plants and to let only the best food animals reproduce. Early scientists learned how plants make seeds, and began deliberately crossing plants to create better crops.

**A researcher at Cornell University examines petri dishes of genetically modified rice plants.**

Recently, though, genetic technology has allowed scientists to develop crops in an entirely new way. Instead of crossing plants many times to get the traits they want, scientists can insert a gene for a desired trait directly into plant cells and grow new plants from these cells. Scientists have used this technique, called genetic engineering, to produce tomatoes that do not rot, corn that makes its own pesticides, and soybeans that are unaffected by herbicides that are sprayed to kill the weeds around them.

Genetic engineering has many benefits for farmers. Crops that resist insect attack may reduce the need for pesticides. Crops that can tolerate being sprayed by herbicides help make weed control much easier. Crops can be engineered to resist disease, ship well, taste better, or have more vitamins.

But there are potential problems as well. Some people oppose any kind of genetic engineering of crops. They believe that scientists understand too little about engineered crops and may not yet understand health risks that could result from eating engineered foods. Some people are concerned that genetically engineered foods may cause allergies that normal foods would not, or that genetically modified foods might not be as nutritious as normal foods. There is also concern that genes inserted into plants to make them resistant to herbicides might be transferred to weeds and other wild plants.

Genetically modified food is not just something to think about for the far-off future. Genetic engineering of food crops is happening today. The work is difficult and the failure rate is high, but success can mean a new and valuable crop. That is why genetic engineering of plant foods is one of the fastest-growing areas of crop science.

Farmers in San Juan Bautista, California, discuss the option of planting genetically modified strawberry plants in order to produce better fruit.

# CHAPTER 1

## DEVELOPING NEW CROPS BY BREEDING

A farmer's success depends on crops. The ideal crop would resist pests, tolerate both drought and heavy rains, and produce large harvests of delicious and nutritious food year after year. No such crop actually exists, but for generations farmers used a technique called mass selection to improve their crops and livestock. Farmers saved seeds from the best food plants for the next year's planting. Farmers also chose their best animals as breeding animals each fall, and slaughtered the rest for the winter's meat. Mass selection gradually led to better plants and animals when done carefully and consistently.

But mass selection was a slow process. Farmers were not always systematic about it, because some farmers preferred to eat the best plants and animals. Some years, too, hunger during an unusually long winter might force farmers to eat some of their saved seeds or good breeding animals. People did not know how inheritance of traits worked, nor did they know how new traits appeared. Farmers had only their own common sense and generations of trial and error to guide them. Yet out of the slow process of mass selection came nearly all of our modern crops.

When humans first learned exactly how plants reproduce, the science of plant breeding was born. Plant breeding allowed scientists to improve crops far more rapidly than mass selection had allowed.

## THE FIRST PLANT BREEDERS

Around 200 B.C., the Greek naturalist Theophrastus described how people in Assyria carried out a religious ceremony in which they pollinated date palms by bringing male and female flowers together.

Yet even though Theophrastus studied plants closely, he did not recognize that other plants besides date palms transferred pollen, nor did he know what pollen did. No one else knew either until 1694, when a German botanist and physician named Rudolph Jacob Camerarius discovered that flowering plants form seeds only when the pollen from one plant landed on the female parts of another. The ancient Assyrian technique for assuring a large date crop, Camerarius discovered, could be used in other crop plants as well.

## EARLY HYBRIDS

Shortly after this discovery, an English nurseryman named Thomas Fairchild produced the first hybrid plant. Known as "Thomas's mule," the plant was a cross between two garden flowers, a clove pink and a sweet William. This new plant began as an accidental cross that Fairchild found among a set of seedlings growing in a border. Although the seeds of the hybrid were sterile, later crosses produced fertile seed, and a hybrid plant known as Fairchild's sweet William was cultivated for many years afterward.

Fairchild's experiments with garden flowers paved the way for experimental plant breeding and the systematic improvement of cultivated plants. In the 1790s, William Rollison founded the first major ornamental plant breeding program, developing Cape Heath shrubs for greenhouse cultivation. By 1840, a new fashion in gardening called for short plants with abundant flowers to be massed in beds. Hybridizers went to work on producing short-stemmed plants with more flowers than leaves.

Ornamental flowers were not the only plants that could be hybridized. Throughout the late 1800s, scientists working for governments and for private companies used hybridization to develop new types of wheat, crossing wheat from the United States with hardy strains from Russia, Turkey, and Ukraine.

**Bees help to fertilize plants by spreading pollen from one flower to another.**

Their experiments led to new varieties that produced larger grains, resisted disease, and did not drop their grains when dry. These new crops boosted the growth of the wheat industry across the Great Plains of North America.

To create new hybrids, breeders planted many seedlings, then selected the ones with the best traits to cross with each other. Thousands of crosses were needed to produce a few useful hybrids. The process relied a good deal on guesswork and luck, because breeders had no idea that inheritance of traits followed predictable patterns. In the middle of the 1800s, however, a Moravian monk named Gregor Mendel made a discovery that later revolutionized plant and animal breeding.

# LUTHER BURBANK, PLANT BREEDER

Among the plant breeders of the 1800s and early 1900s, one of the most famous and productive was Luther Burbank. Born in Lancaster, Massachusetts, in 1849, Burbank later moved to Santa Rosa, California, where he lived for more than fifty years. Burbank was an enthusiastic horticulturist, and made tens of thousands of crosses between plants in his search for new food crops and garden flowers. Of those crosses, hundreds turned out to be successful. Burbank created the Shasta daisy, which is a favorite garden flower today. His improved Burbank July Elberta

Luther Burbank was one of the most famous plant breeders of his day.

peach gave farmers earlier peach crops. Burbank also produced the giant garlic known as elephant garlic, which is still a favorite of gardeners.

Burbank's daisy varieties remain popular today.

One of his most important new crops was the Burbank potato. His work was inspired by the massive potato crop failure in Ireland that led to the great potato famine of the 1840s. The Irish had been growing a single variety of potato throughout their country, and when a series of wet, rainy years caused fungal diseases to flourish, the potato crops were devastated. Burbank thought he could help prevent such famines by developing disease-resistant potatoes that also produced better crops. He started with seeds of a potato variety known as Early Rose and found that one of the seedlings produced more potatoes than the others. The potatoes also resisted rot and stored well. Careful cultivation produced more potato plants that retained the same traits. The potato was an instant success, and one variety developed from it, the Russet Burbank, became the basis for the Idaho and Washington potato industry.

# THE DISCOVERY OF GENETICS

Mendel was a farmer's son who was fascinated by science. He had many questions about the heredity of food crops. Mendel joined a monastery run by an abbot who was also interested in science. With the encouragement of the abbot, Mendel went to the University of Vienna to study science, then began his work on the inheritance of traits in pea plants.

Peas were an ideal plant for studying heredity. Mendel could grow at least two crops each year. Pea vines could be trained upright, so he could grow many plants in a small space. And pea plants had several distinct traits, such as flower color, pea color, height, and pea shape, that could be studied easily.

Mendel spent nine years studying peas. He carefully recorded the traits in each parent plant and in their offspring. After several harvests, he discovered that traits did not just blend together as most people assumed. The traits he was studying remained distinct, and were passed on in predictable patterns. Mendel spent years testing his ideas, making predictions and then recording the outcomes, to be sure of his results.

When Mendel presented his findings at a meeting of a natural history society in 1865, few people were able to understand the importance of his work. It was not until 1900, after Mendel's death, that scientists who were trying themselves to discover the laws of heredity rediscovered Mendel's published papers and found out that much of their work had already been done.

# ALTERING PLANT GENETICS

The rediscovery of Mendel's work led scientists to theorize that traits were carried by some sort of particle, which they called "genes." Variation of traits was critical to crop development.

By experimenting with pea plants, Gregor Mendel discovered the laws of genetics.

# WHAT MENDEL DISCOVERED

Gregor Mendel, the monk who discovered heredity, used mathematical patterns to show that inheritance can be predicted. Mendel started with pea plants that were pure breeding—that is, they produced certain traits but not others. For example, to study flower color, he chose plants that always produced offspring with purple flowers and crossed them with plants from strains that always produced white flowers. The offspring from the cross all had purple flowers. But when he crossed two of the offspring, both traits turned up in the next generation. Three-quarters of

Mendel used pea plants in his research to understand inherited traits.

the plants had purple flowers, and one-quarter had white flowers.

Mendel called these two flower traits "dominant" and "recessive." Mendel said

In pea plants, purple flowers represent a dominant inherited trait.

that the dominant trait, purple, was carried on one hereditary particle. And the recessive trait, white, was carried on another hereditary particle. In the pure-breeding plants, the purple-flowered plants had only the purple particles, and the white plants had only the white particles. Their offspring inherited the purple particle from the purple parent and the white particle from the white parent, but their flowers were purple, not a blend of purple and white. This showed that traits were transmitted on particles of some kind, not through the blending of chemicals as many people thought.

When the offspring crossed, each of them could give either the white particle or the purple particle to their offspring. Some of their offspring got two purple particles, some got a purple and a white, and some got two whites. Only the ones with two white particles showed the recessive white flower trait. Later, these particles would be known as genes.

Scientists work with seeds and plants on an experimental farm. The seven-foot pipe in the middle of the planter releases radiation into the soil.

Different varieties of plants had different traits, and therefore they had different genes that could be mixed with those of other plants to create new varieties.

In the 1920s, scientists found that not only could they remix traits through hybridization, but they could also produce completely new traits by exposing plants to radiation. Scientists hypothesized that radiation somehow altered the genes, though they did not know how. At that time, no one knew exactly what genes were nor what they were made of.

New traits produced by radiation were called mutations. One of the first commercial crop plants produced by this new method was an oat, the Alamo-X. The oat had a mutation caused by exposure to X-rays that helped the plant resist blight. Two grain crops, Calrose 76 rice and Golden Promise barley, were produced using gamma rays. In both cases, the mutations produced shorter, sturdier plants with higher yields.

Still, the radiation method was not as efficient as scientists wanted. They had to irradiate thousands of seeds and observe thousands of plants before they found one mutation that was useful. Before they

**This highly magnified photo shows plant cells and their nuclei. Mutations occur in the nuclei of cells, where DNA is stored.**

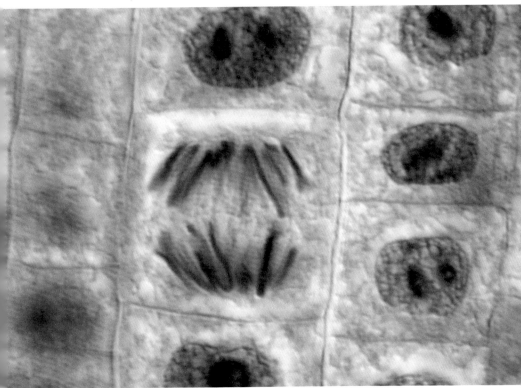

# The DNA Molecule

The DNA molecule is a spiraling strand of chemicals inside every cell of an animal or a plant. The rungs of this ladderlike structure are made of base pairs: adenine, thymine, guanine, and cytosine. The arrangement of these base pairs forms a code that determines all of a plant or animal's inherited traits.

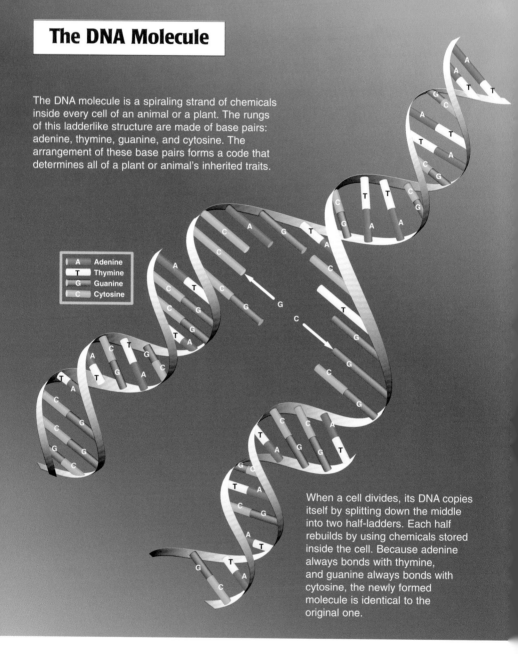

| A | Adenine |
| T | Thymine |
| G | Guanine |
| C | Cytosine |

When a cell divides, its DNA copies itself by splitting down the middle into two half-ladders. Each half rebuilds by using chemicals stored inside the cell. Because adenine always bonds with thymine, and guanine always bonds with cytosine, the newly formed molecule is identical to the original one.

could develop more efficient methods of crop development, scientists needed to understand exactly what genes were and how they worked. The search for the gene led to the discovery of DNA, which is the material that genes are made of.

# THE DISCOVERY OF DNA

Ever since the middle of the 1800s, scientists have known that when cells prepare to divide, structures called chromosomes appear inside the cell. Scientists analyzed chromosomes and found that they were made of materials called nucleic acids. Some scientists speculated that nucleic acids might have something to do with heredity.

In 1943, American scientist Oswald Avery showed that a certain nucleic acid called deoxyribonucleic acid (DNA) carried information that created traits in organisms. Avery believed that DNA must be the material that chromosomes (and the genes they carry) are made of, though he did not understand how DNA worked.

Scientists knew that DNA contained four substances called bases: adenine, thymine, cytosine, and guanine. DNA also contained sugar, as well as phosphorous compounds called phosphates. In 1950, a biochemist named Erwin Chargaff discovered that adenine and thymine were always present in equal amounts in DNA. So were cytosine and guanine. This added one more clue to the DNA puzzle: The bases must be arranged in some orderly way.

Two teams of scientists in England worked to discover the structure of DNA. Rosalind Franklin and Maurice Wilkins crystalized DNA and then took X-ray pictures. Analysis of the pictures led them to believe that DNA had a spiral structure. James Watson and Francis Crick looked at Franklin and Wilkin's X-ray pictures and also thought that the structure looked something like a twisted ladder. They called this shape a double helix. Chargaff's discovery led Watson and Crick to believe that the rungs of the ladder were formed by pairs of bases. Adenine must be paired up with thymine, since both were found in equal amounts. For that same reason, cytosine and guanine must also be paired. The sides of the ladder were made of the sugars and phosphates. Watson and Crick used their ideas to build a model of the DNA molecule. They showed that the ladderlike molecule could duplicate itself by splitting apart down the middle as the pairs of bases separated from one another. More bases bonded with the unzipped ladder to form two new ladders that were identical to the original.

Once scientists knew the structure of DNA, they were able to discover exactly how genes worked. Over time, scientists found that the order of the bases in DNA was important, because the bases formed a code that contained the instructions for making protein molecules. Proteins may form structures, such as hair or fingernails. They may be enzymes that carry out many cell processes. They may be pigments that give skin or eyes their color. Proteins are the basis of all traits that can be inherited.

Once the genetic code was understood, scientists began identifying specific genes that held the instructions for specific traits. Over many years, scientists discovered how to isolate a gene from the rest of the organism's DNA. These discoveries paved the way to a new method of producing improved crops: genetic engineering.

# CHAPTER 2

## GENETIC ENGINEERING OF FOOD CROPS

Mass selection allowed farmers to develop crops with the best traits available, but the farmers could not produce new traits themselves. The discovery of radiation allowed scientists to mutate crop plants in hopes of producing new, desirable traits, but out of thousands of mutations, only one or two might be useful. Genetic engineering allows new transgenic crops to be developed much faster. Genetic engineering is the science of finding and isolating a gene for a desired trait and then deliberately putting that gene into plant cells.

**A researcher examines the flowers of a crop that has been genetically engineered to resist weed killer.**

A scientist works on isolating a specific gene from eastern gamagrass that might be used to create a tastier variety of corn.

# NEW GENES AND TRANSGENES

Before scientists can genetically engineer a new food crop, they have to find the genes for the traits that they want. Chromosomes may contain thousands of genes, so finding a specific gene is difficult. Once scientists find a gene, they must isolate the DNA that makes up the gene and turn it into a transgene. A transgene is a sequence of DNA that contains several parts: a promoter sequence, the gene itself, a termination sequence, and, sometimes, a marker gene.

A promoter sequence is a piece of DNA that controls exactly when a needed gene is copied and its information used to make a protein. The gene copying process takes place inside the nucleus of the cell, where DNA is located. When a certain protein is needed, the cell sends a chemical signal to the nucleus. The promoter sequence on the gene for that protein then allows enzymes to unzip the part of the DNA that contains the gene. The enzymes match bases to the unzipped part of the DNA and make a copy of the gene. The copy moves out of the nucleus and is picked up by a structure called a ribosome, which reads the sequence of bases and uses the information to assemble the protein from smaller molecules called amino acids.

After the promoter sequence comes the gene itself. This is a single gene that codes for a specific trait that has been isolated from another plant or from an animal.

Next comes a termination sequence. This is a piece of DNA that contains a certain sequence of bases that the gene-copying enzyme recognizes as a code for "stop." The enzymes stop copying the gene at that point.

A transgene may also have a marker gene attached to it. A marker gene is useful if the desired gene is for a trait that is not obvious in a young plant. For example, after scientists insert a transgene that makes a plant resistant to a disease, they want to know which plants have successfully taken up the gene into their

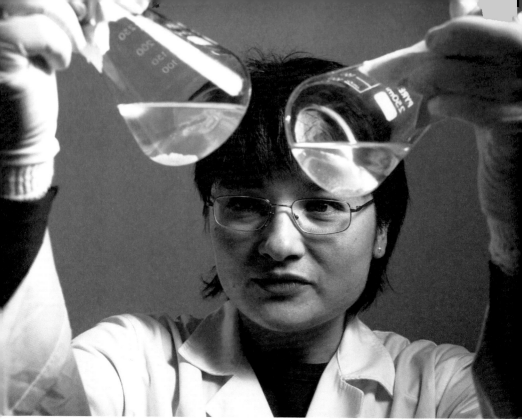

**A scientist involved in transgenic plant research in Italy examines flasks of cultured rice cells.**

own DNA. Disease resistance is not something that they can see just by looking at the plant. A marker gene for a trait that is easy to detect is attached to the transgene before treating the plant. One of the first marker genes was a gene that caused jellyfish to glow in the dark. After inserting the transgene with the marker gene into plant cells, scientists looked for young plants that glowed in the dark. These were the plants that had successfully taken up the transgene. Later on, scientists found a gene that made plants resistant to herbicides. This was useful by itself as an inserted gene in crop plants, but it was also useful as a marker gene. Genetically altered plants could be sprayed with herbicide. Those that survived were the ones that had taken up the whole transgene, including the herbicide-resistant marker gene.

After the transgene is constructed, the next step is to get the transgene into plant cells. Scientists start with plant cell tissue cultures. These are clusters of plant cells grown in petri dishes. There are several methods for getting genes into these cells. The two most common ones are the bacterial method and the gene gun.

## INSERTING GENES

The bacterial method takes advantage of the way in which some bacteria naturally cause diseases in plants. One species of soil bacteria, *Agrobacterium tumefaciens*, injects its own DNA into the cells of the

*Agrobacterium tumefaciens* **bacteria (red) grow on the surface of a tobacco plant cell in this highly magnified image**

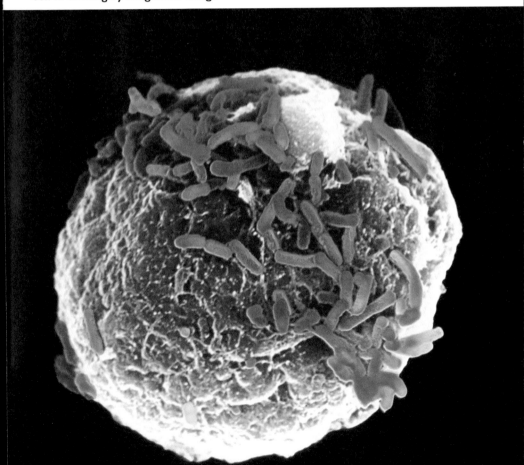

host plant. This forces the plant cell to make a tumor, where more bacteria can grow. Crop scientists use this bacteria by removing its DNA and replacing it with a transgene. The modified bacteria are then added to a plant cell culture. The bacteria attach themselves to plant cells and inject the transgene. The method is very useful because each bacterium appears to attack an individual cell, so only one copy of the desired transgene is inserted into a cell. The bacterial method has been most successful with broadleaf crops such as potatoes. Scientists have had less success in using the bacterial method with other crops, such as grains. For these, they use a gene gun.

A gene gun gets its name from the way in which it literally shoots DNA into cells. To use a gene gun, scientists coat microscopic particles of gold or tungsten with the transgenes they

**A biologist uses an apparatus called a gene gun to insert transgenes into plant cells cultured in petri dishes.**

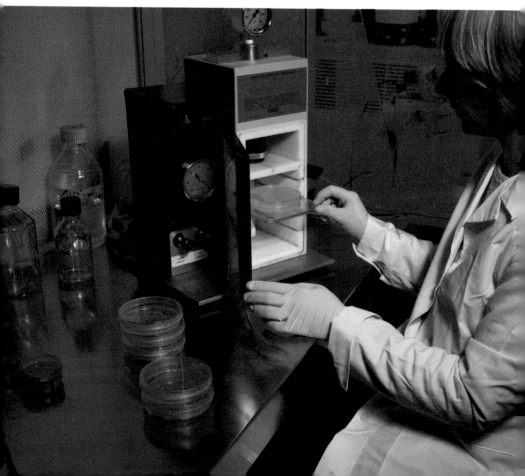

**Transgenic rice rich in vitamin A grows in test tubes in an experiment conducted in the Philippines.**

want to insert. The particles are attached to a plastic bullet, which is loaded into the gun, and the gun is aimed at a tissue culture. An explosive charge drives the bullet forward, but it does not leave the barrel. Instead, the impact of the bullet against the end of the barrel sends the DNA-coated particles flying into the tissue culture. Some of the particles enter the nuclei of the cells, and the transgenes may be taken up into the cells' DNA.

Unlike the bacterial method, which usually inserts only a single copy of the transgene into a cell, the gene gun may shoot many copies of the transgene into any one cell. If too many are taken up into the plants' DNA, the cell may not function properly. However, the gene gun method has been successful with rice, corn, and other grasslike crops.

After the transgenes have been inserted into the cells and have been taken up into the cells' DNA, the tissue is then transferred to new culture dishes with a different material to grow on. This

# VIRUSES AS GENE CARRIERS

To insert DNA into cells, scientists may use a virus as a gene carrier. Viruses are nonliving particles made up of a molecule of DNA or a similar genetic material, ribonucleic acid (RNA), with a coat of proteins wrapped around it. Viruses cannot reproduce on their own, but they do have a way of producing more viruses. To do this, they hijack the DNA of living cells.

When a virus attacks a cell, it injects its own DNA or RNA, along with some enzymes, into the host cell. The genetic material and the enzymes make their way to the host cell's nucleus, where they force the cell to use the genetic material to make more viral protein coats. The enzymes make more copies of the viral DNA, and assemble the new viruses. Eventually the host cell is filled with viruses and bursts open. The released viruses attack other cells, spreading the viral disease.

Scientists reasoned that if viruses could inject viral DNA into cells, viruses could be forced to inject other DNA, too. They discovered ways to remove the viral DNA and insert a desired gene, such as a transgene, into the empty protein coat. When the modified viruses attack a cell, they inject the cell with the DNA for the desired gene. The cell will not make more viruses, however, because no viral DNA has been injected. Instead, if all goes well, the inserted gene will be incorporated into the host cell's DNA and produce the protein that scientists want.

These harmless BK virus particles are commonly found in the urinary tracts of adults.

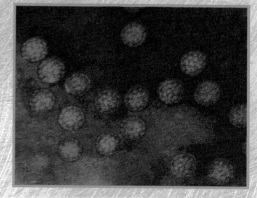

material may allow only the transformed cells to survive, due to the action of the marker gene. For example, if the marker was a gene for resistance to an herbicide, the growth medium would have herbicide mixed into it.

The genetically transformed cells are then treated with plant hormones to cause the cells to grow into new plants. These plants are grown to maturity and allowed to make seed. Offspring grown from these seeds are tested to see if they, too, contain working copies of the inserted genes.

## TESTING NEW PLANTS

Finally, each transgenic plant is tested to see how stable the new genes are in that particular plant. Scientists want to be sure that the new genes will be passed on through many generations and will still continue to function.

Even after passing these tests, a genetically transformed plant may not become a crop immediately, because the varieties of plants that scientists are able to transform successfully in the laboratory may not be the most useful crop varieties. The transformed plants may instead be crossed with existing plants that have other desirable traits. The resulting hybrids may be the seeds or plants that finally go to market.

## GENETICALLY ENGINEERED CROPS TODAY

Creating genetically engineered crops is a long and difficult process. However, the process has produced a number of useful crops. Many varieties of genetically engineered cotton, soybeans, corn, tomatoes, potatoes, and other crops are currently being grown in the United States. Some of these have been approved as food crops. Others have been approved for fibers, animal feed, and other uses.

**An American farmer displays a handful of genetically modified soybeans.**

One of the most common uses of genetic engineering is to make crop plants resistant to herbicides. This allows farmers to spray their fields with weed-killing chemicals without harming the crops.

Another common use of genetic engineering is to insert a gene for the Bt toxin into crop plants. Bt toxin is a chemical produced by a soil bacteria called *Bacillus thuringiensis*. This chemical is poisonous to caterpillars and grubs, but it is believed to be harmless to mammals, including humans. Genetically engineered plants that produce this toxin are resistant to pest damage. Farmers do not have to spray these crops with pesticides, many of which are toxic to humans and other mammals.

While genetic engineering may combat pests and solve many problems in agriculture, it is not without risks. Concern over risks may affect the future of genetic engineering.

# CHAPTER 3

## THE FUTURE OF GENETICALLY MODIFIED CROPS

Transgenic technologies are currently being applied to nearly every major food crop. In many cases, scientists are working simply to understand the genetics of a particular crop. In other research programs, crop scientists are attempting to increase disease resistance, decrease ripening time, improve nutrition, or improve the overall quality of various crops.

## CROPS CURRENTLY UNDER RESEARCH

One area of current research is the development of salt-tolerant plants. Salty soil is a growing problem in dry areas of the world. Crop irrigation may increase the salt content of the soil, because the water used for irrigation is often high in minerals. The water evaporates or is taken up by the plants, but the minerals, including salt, remain in the soil. Salty soils damage many crop plants, reducing the value of the land. Salt-tolerant crops could help farmers reclaim some of this land for production, though improved irrigation methods may still be necessary to prevent the soil from accumulating even more salt. One variety of salt-tolerant tomato is nearly ready for commercial production.

Besides developing crops with traits that benefit farmers, scientists are also working on crops with traits that consumers want. One example is a new type of coffee currently under research. Millions of people around the world enjoy drinking coffee but do not want to consume caffeine. Coffee can be decaffeinated, but some decaffeinating processes use chemicals that people would rather avoid, and some people believe the decaffeinating process decreases the flavor of coffee. Crop scientists are working on turning off the genes that create caffeine in coffee plants, in hopes of producing a full-flavored caffeine-free coffee.

Scientists are also experimenting with genes that control fruit ripening. Plants have genes that cause fruits to develop and ripen. Part of the ripening process includes softening of cell walls, which makes the fruits softer and more prone to bruising. By modifying the ripening genes, scientists may be able to delay the softening process in fruits, allowing fruits to be shipped with less damage. One early experiment in this technology was the Flavr Savr tomato. However, the transformed tomato lacked flavor, so it did not catch on with consumers.

Improvement of food plants is one line of research in genetic engineering, but there are other areas of research as well. For example, scientists are working on inserting genes into crop plants to make the plants produce certain medicines and other pharmaceuticals. This area of research is known as bio-pharming.

**A researcher studies tomato seedlings in a lab in an effort to understand the genetic factors that control ripening.**

Although genetically engineered *Flavr Savr* tomatoes (upper right) last longer than the natural variety, they are less flavorful.

## TRANSGENIC CROPS TO IMPROVE HEALTH

Bio-pharming may help developing nations that are troubled by diseases. Although vaccines may exist for these diseases, they are often expensive. Developing nations may not have the money to buy vaccines, store them properly, and distribute them to the people who need them. Several research teams are working on genetically engineering bananas, which are widely grown in the tropics, to produce proteins found in cholera, hepatitis B, and other diseases that plague poor tropical nations. Bananas altered in this way could serve as inexpensive vaccines that create immunity to these illnesses in the people who eat them, thereby saving thousands of lives.

Another bio-pharming project is the improvement of infant formula. Doctors have long known that breast-fed infants receive

**A worker sorts bananas at a market in Panama. Genetically engineered bananas may some day provide immunity against cholera and hepatitis B.**

enzymes from their mothers' milk. These proteins help improve the child's immune system. Bottle-fed babies do not have this advantage. To solve this problem, researchers have isolated two proteins in human milk that are key enzymes in helping support an infant's immune system. The genes for these two enzymes have been inserted into rice, which is often used to make infant formula. This is an inexpensive way to produce these human proteins and may provide bottle-fed babies with the immune support they need.

Besides producing medicines in plants, crop scientists are also working on improving the vitamin content of some plants. One example is a type of rice called Golden Rice. In some parts of the world, people eat very little besides rice. Their monotonous diets can cause deficiencies in important vitamins, including vitamin A. People who do not get enough vitamin A develop vision problems, since it is used to produce light-sensitive pigments in the retina of the eye. To help solve this problem, crop scientists in Switzerland inserted three genes into rice: two from daffodils and one from a bacteria. These genes work together to cause the rice to produce beta carotene, a yellow plant pigment that the human body uses to make vitamin A. This new type of rice is still being crossed with high-producing strains in hopes of developing a productive, vitamin-rich rice crop.

Though genetically engineered crops under development may help solve problems in farming, food distribution, and health, many people are concerned about risks associated with these crops. They are worried that scientists do not know enough about the effects of altering the genes of a food crop.

**The Golden Rice shown here may help supplement diets deficient in vitamin A, essential for healthy vision.**

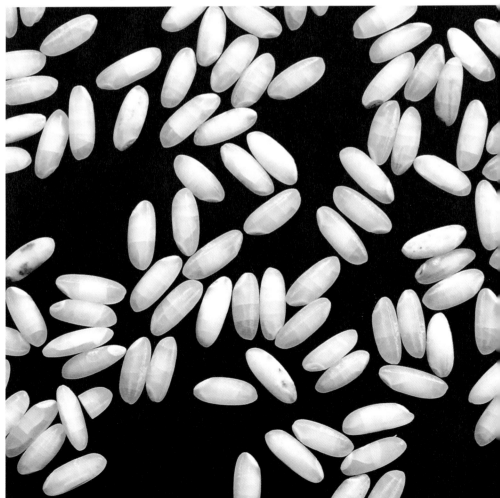

Concerned about the potential health risks of genetically engineered food, a group of protestors demonstrates outside the Food and Drug Administration in Chicago.

## POTENTIAL RISKS TO HUMAN HEALTH

People who have food allergies are very concerned about their food supply, particularly about the proteins in their food. Most allergies are caused by the human immune system reacting to specific proteins, such as those found in nuts, fish, or eggs. In some cases, eating a certain protein can cause a dangerous or even fatal immune reaction. If a protein from nuts, for example, were inserted into a common food crop, people who are allergic to nuts would not be able to eat the altered food.

One type of soybean was developed using a gene from Brazil nuts to improve the quality of protein in the beans. While the crop was intended only for animal feed, the developers knew that during harvest and storage of soybeans, it is hard to keep soybeans for human food separated from those meant for livestock. They decided to test the soybeans to see if they were a potential danger. In laboratory tests, the scientists found that people who were allergic to Brazil nuts also had a reaction to the soybeans. Because of this, the company that produced the soybean stopped further development and never grew the beans commercially.

## CROSS-POLLINATION

Besides allergic reactions, there is concern that crops that are engineered to produce pharmaceuticals may cross-pollinate with plants intended for food. Most biopharmaceutical crops are grown in small, isolated fields, well apart from farms that grow crops for human food. However, grain crops such as corn, wheat, and rice produce large quantities of pollen, which is carried by the wind to other plants. While most of the pollen travels only a small distance, some may get blown far enough to pollinate plants in a field of grain intended for human consumption. Any seeds produced by this cross-pollination would contain the gene for the pharmaceutical product. If some of that seed happens to be saved for next year's crop, the new crop would produce the pharmaceutical.

Avoiding genetically engineering wind-pollinated plants would not necessarily reduce the risk of cross-pollination, either. Many crop plants rely on bees to carry pollen from one plant to another. Bees tend to forage for pollen near their hives, but they may also fly far from the hive to look for other pollen sources, and so may carry pollen from field to field.

# TRANSGENIC ANIMAL RESEARCH

One area where transgenic science has not been successful has been in producing genetically modified livestock for food production. Some work has been done on producing pigs engineered to grow faster, which would increase meat production, but the genetics of growth are complicated. Attempts to increase growth rate produced unhealthy animals and were discontinued.

Animals have, however, been used successfully in bio-pharming, just as transgenic plants have been. For example, goats have been genetically altered to produce pharmaceutical proteins in their milk. Milk already contains a protein called casein. Scientists can replace the gene for casein with a transgene for a pharmaceutical protein without harming the goat. The proteins can be purified from the milk and used as medicines,

Scientists can genetically alter a goat so that proteins in its milk can be used to make medicine for people without harming the goat.

Chickens can be genetically altered to produce a bacteria-killing enzyme called lysozyme.

and the remaining whey and milk solids can be used for other things, such as animal feed. The milk is not used for human food. One pharmaceutical protein that has been produced in this way is an anticlotting blood protein that is useful in surgeries and for treating people who have had heart attacks.

Another area of animal research that shows promise is current work on chickens to increase production of an enzyme called lysozyme. This enzyme already occurs naturally in egg whites, where it helps kill bacteria in eggs soon after they are laid. However, eggs only make small quantities of lysozyme naturally. Conditions in large-scale egg production farms, where hens live close together, allow bacteria to multiply and spread easily. Increasing the amount of lysozyme in egg whites may help overcome this problem and make eggs safer. Scientists are also researching ways to isolate the bacteria-fighting enzyme from egg whites and use it as an antibiotic or a food preservative.

Potential risks of genetically engineered crops to human health are only one area of concern. Some people question whether plants that are altered to resist pesticides and diseases are truly effective.

# EFFECTIVENESS ISSUES

Cotton and corn are two crops that have been genetically altered with the gene for making the Bt toxin, making them poisonous to caterpillars and grubs, but not to humans. In theory, this should allow farmers to use far fewer pesticides on their crops. However, critics of genetic engineering question whether farmers who are growing these crops are really spraying less.

**Genetically engineered crops may reduce some of the need for pesticides, but farmers may still be using pesticides to fully control pests.**

In recent years, the Environmental Protection Agency has conducted a series of investigations to determine if planting Bt-altered corn helps reduce the use of pesticides.

The United States Department of Agriculture (USDA) investigated pesticide use in Bt-altered cotton in six states where cotton is a major crop. They compared farmers' records of pesticide use before and after they planted Bt-altered cotton. At the end of the study, the USDA reported that the amount of pesticide that farmers used had dropped about 14 percent. Although significant, this was not as great a reduction as crop scientists had hoped.

The Environmental Protection Agency (EPA) carried out a series of similar investigations on Bt-altered corn. Investigating pesticide resistance in corn is more difficult than studying cotton, because there are many more insect pests that can attack corn, and destroying one kind of insect may allow another kind to flourish. While the EPA found small reductions in pesticide use among farmers who grew Bt-altered corn, ranging from 1 percent to 8 percent, the data were not conclusive. Critics of genetic engineering say that these small reductions in pesticide use are not enough to justify the costs and risks of developing Bt-altered corn.

## THE FUTURE OF ALTERED CROPS

Though some people have deep concerns about genetically modified foods, such foods are in stores now and more are under development. Scientists, policy makers, and public opinion will work, sometimes together, sometimes at odds with one another, in shaping the future of genetically modified foods.

# GLOSSARY

**bacteria:** One-celled organisms that lack a nucleus.

**cell:** The basic unit of all living organisms. Each cell has a membrane as an outer boundary, as well as many internal parts that carry out the processes of life.

**cross:** When two organisms of different species, or two varieties of the same species, interbreed.

**DNA:** Deoxyribonucleic acid, the molecule that contains the genetic code.

**gene:** A section of DNA that directs the formation of a particular protein.

**gene gun:** A mechanism that uses an explosive charge to insert genes into cells.

**genetic engineering:** Changing the genetic makeup of an organism by inserting new genes.

**heredity:** The process by which traits are passed on from one generation to the next.

**hybrid:** The offspring of a cross between parents of two closely related species or two varieties of the same species.

**mutation:** A change to the DNA of an organism. Some factors that cause mutations include exposure to radiation and strong chemicals.

**nucleus:** A structure inside of cells that contains the DNA.

**pollination:** When pollen lands on the female parts of a flower and fertlizes the structure that becomes the seed.

**resistance:** The ability to resist the effects of a poison or a disease.

**tissue:** A group of cells that have similar structure and function, and often work together.

**transgene:** A structure manufactured from DNA, made up of a start sequence, a gene to be inserted into an organism, a marker gene, and a stop sequence.

**virus:** A nonliving structure made up of a piece of genetic material inside a protein coat.

# FOR FURTHER INFORMATION

*Books*

Holly Cefrey, *Cloning and Genetic Engineering*. Danbury, CT: Children's Press, 2002.

Karen Judson, *Genetic Engineering: Debating the Benefits and Concerns* (Issues in Focus series). Berkeley Heights, NJ: Enslow, 2001.

Lisa Yount, *Gene Therapy*. San Diego: Lucent, 2002.

*Periodicals*

David Bjerklie, "Trouble on the Table?" *Time for Kids*, March 2000.

Barbara Eaglesham, "Taking Sides in a 'Food Fight,'" *Odyssey*, October 2000.

George Erdosh, "Designer Foods," *Odyssey*, October 2000.

Cindy Maynard, "Biotech at the Table," *Current Health 2*, November 2000.


*Web Sites*

**Cloning Plants by Tissue Culture** (www.jmu.edu/biology/biofac/facfro/cloning/cloning.html). This site gives an overview, with photos, of how plant cells are cultured.

**Plant Breeding: Creating New Plants, British Broadcasting Corporation** (www.bbc.co.uk/science/genes/gene-safari/breeding-zone/plant.shtml). This site gives an overview of plant breeding and genetic engineering of new crops.

**Transgenic Crops: An Introduction and Resource Guide, University of Colorado** (www.colostate.edu/programs/lifesciences/TransgenicCrops/index.html). This Web site provides general information on transgenic crops, including how crops are transformed. The site also discusses issues and concerns about transgenic crops.

# INDEX

# ABOUT THE AUTHOR

Karen E. Bledsoe has written or cowritten 30 nonfiction books for children, including many books on science and technology. Her articles have appeared in such magazines as *Odyssey* and *Appleseeds*. She teaches biology at Linn-Benton Community College in Albany, Oregon, and is working on a PhD in science education. She is a Fellow with the Oregon Writing Project at Willamette University in Oregon. When she is not writing or teaching, she enjoys hiking and gardening.